A special thanks
to everyone
who has helped make
Know Yourself
what it is today.

Dear Reader

Knowing yourself is truly the beginning of all wisdom. We give young learners the building blocks they need to start their unique journey of self-discovery: an understanding of human anatomy — literally how we are put together. Knowledge of one's own human body is an empowering context on which anyone can build.

Learning about the body and mind at a young age sets the foundation for honoring one's physical form, develops self-esteem and self-confidence, and begins the discovery of who we are meant to be in this world.

Now that's real power.

The Know Yourself Team

ADVENTURE 8
Quick-Start Guide
Hello Know Yourselfers!

Follow these steps to start a new journey and explore the endocrine system. Have fun on this quest and remember - keep the information flowing!

1 **Grab a baseball bat and a guante de béisbol** [baseball glove] **, we're going to Cuba!**

Locate Cuba on your atlas, or find an online map of the world.

2 **Read Time Skaters Adventure 8.**

Pinky and Stokely take an unexpected walk to 1954! It's a battle of brain vs. brawn as they score some new clues about their trips through time.

3 **Get equipped!**

Gather your supplies and prepare for your activities. You'll need all the right stuff to bring it home.

III

Table of Contents

QUICK-START GUIDE III

    Hello Adventurer! 1

TIME SKATERS ADVENTURE 8 4

LEARNING CALENDAR 40

    Home Inventory Checklist 42

PART 1 • KNOW YOUR HISTORY 44

LEARN:
- **Know Your History - Cuba** 44
- **Know Your Cuban Baseball** 46

ACTIVITIES:
- Make Baseball's Secret Language Your Own .. 48
- Know Your Son Cubano 50
- Make Your Own Clave Rhythm 52
- Know Your Cuban Poetry 56
- Hidden Message Poems 58
- Know More Spanish 60
- Evolution of the Revolution Crossword 62
- Cuban Comprehension 64

PART 2 • KNOW YOUR ENDOCRINE SYSTEM .. 67

LEARN: Know your Endocrine System 67

Know your Pancreas 72

ACTIVITIES:
- Show Me the Sugar 74
- Adrenal Aegis 78
- Notice Your Adrenal-ine 80
- Butterfly Beat 82
- Matchmaker 86
- Chemical Constitution 88
- Endocrine Expertise 90
- Talk About a Team 92

PART 3 • KNOW YOUR APPETITE 84

RECIPES:
- Cuban Style Black Beans 86
- Mojo Sauce 89

Thoughts for Young Chefs 92

PART 4 • CONCLUSION 94

- In it to Win it 99

Further Reading 99

Hello Adventurer!

Welcome to Adventure 8!

Hello, Adventurer! In this workbook, you will learn about Havana, Cuba and your body's Endocrine System. There will be information to read, activities to complete, and quizzes to take when you are ready to challenge yourself! Take your time along the way - spend as much or as little time as you like on each activity.

Good luck, and have fun!

Destination: Havana, Cuba!

THE TIME TRAVEL CLOCK READS 1954

ADVENTURE 8

Learn about the Endocrine System.

There are 9 glands in the Endocrine team lineup, producing hormones that zip around your body, communicating through special channels.

We hope you enjoy this Adventure, but remember, to become an endocrinologist requires further study!

VISIT:
1954 Havana, Cuba

In the Caribbean Sea, this island republic is full of music, poetry, and baseball.

Meet Pepe, Jaime, and other athletes learning to work together.

THE ENDOCRINE SYSTEM 1

Enter this portal for….

Time Skaters Adventure 8
Grand Slam!

THE ENDOCRINE SYSTEM

THE ENDOCRINE SYSTEM

THE ENDOCRINE SYSTEM

THE ENDOCRINE SYSTEM

THE ENDOCRINE SYSTEM

FUN-damental Science

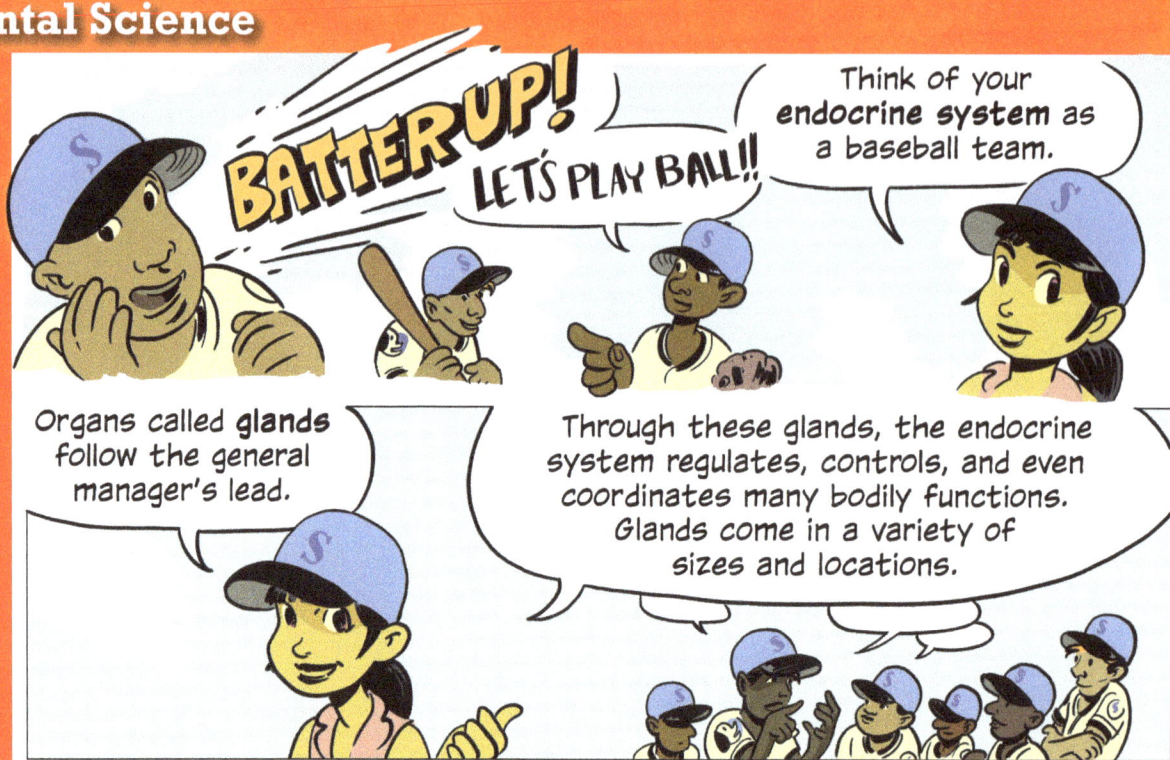

BATTER UP! LET'S PLAY BALL!!

Think of your **endocrine system** as a baseball team.

Organs called **glands** follow the general manager's lead.

Through these glands, the endocrine system regulates, controls, and even coordinates many bodily functions. Glands come in a variety of sizes and locations.

First Pitch: Hormones Set Things in Motion! Each gland produces **hormones**, which are chemical signals, or messages, that travel in the bloodstream and bodily fluids.

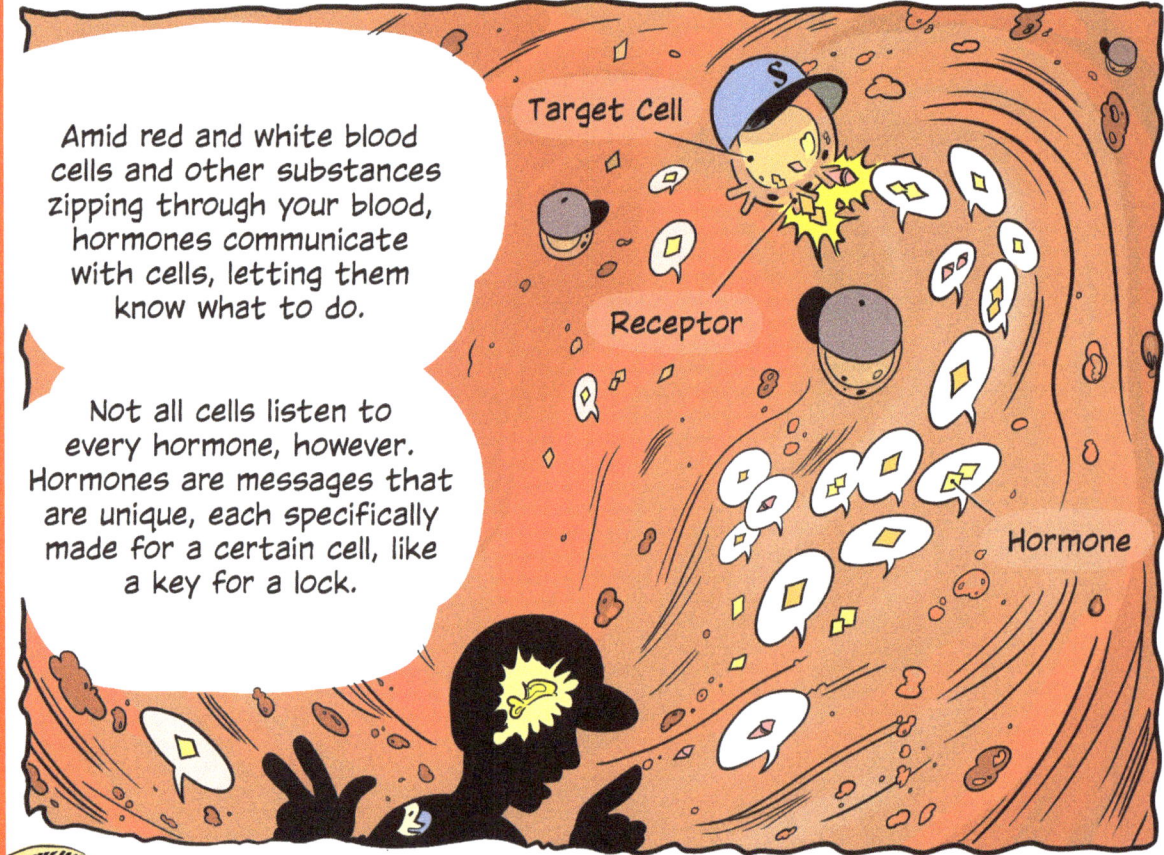

Amid red and white blood cells and other substances zipping through your blood, hormones communicate with cells, letting them know what to do.

Not all cells listen to every hormone, however. Hormones are messages that are unique, each specifically made for a certain cell, like a key for a lock.

FAST FACT Say it like this: endocrine "end-o-krin"

Besides endocrine glands, hormones can come from many other cells in the body, like those in your heart and stomach.

12   ADVENTURE 8

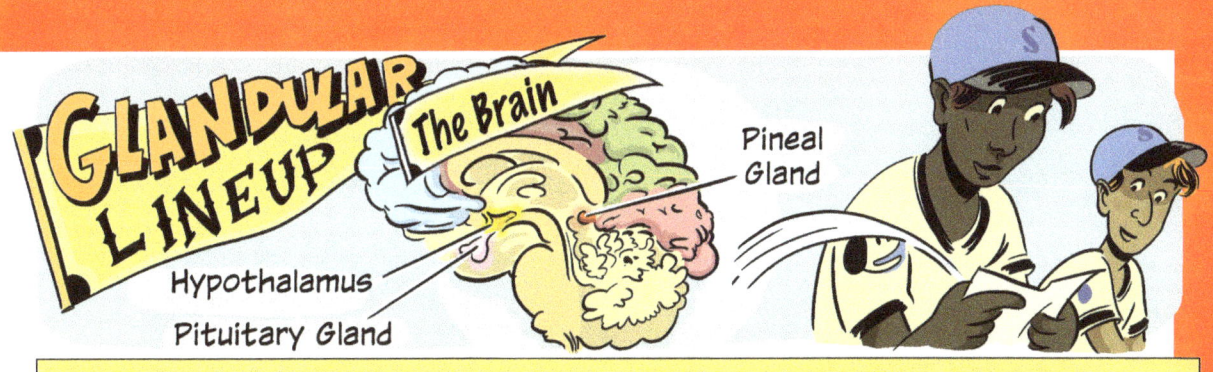

GLANDULAR LINEUP

The Brain
- Hypothalamus
- Pituitary Gland
- Pineal Gland

Some of the glands in the endocrine system are the **hypothalamus, pituitary, thyroid, parathyroid,** and **adrenal glands.** (There is even the **thymus gland,** which helps immunity until you're in your early teens, then dissolves into fat.)

Like any team, your endocrine system has lots of parts.

Team Endocrine also needs a general manager to direct the overall distribution of messages to the players—or in the body's case, to cells, organs, and tissues.

General Manager **Hypothalamus** processes information from the brain and creates instructions to tell glands when to start or stop producing hormones.

Meanwhile, the much smaller but very important **pituitary** gland is the Assistant Manager, connected to the base of the hypothalamus.

It has two different lobes, posterior and anterior. Posterior stores hormones for the hypothalamus. Anterior produces growth hormone, the "feel good" hormones, endorphins, and other hormones that can turn glands on and off!

Just like Jaime's younger brother Pepe influences the other Sabuesos, the pea-sized pituitary has an enormous effect on the other glands. In fact, it is sometimes called "the master gland."

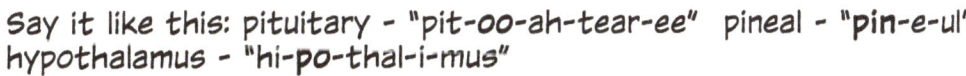

Say it like this: pituitary - "pit-**oo**-ah-tear-ee" pineal - "**pin**-e-ul"
hypothalamus - "hi-**po**-thal-i-mus"

Together these two glands are called the **hypothalamus-pituitary complex!**

FAST FACT

THE ENDOCRINE SYSTEM

THE ENDOCRINE SYSTEM

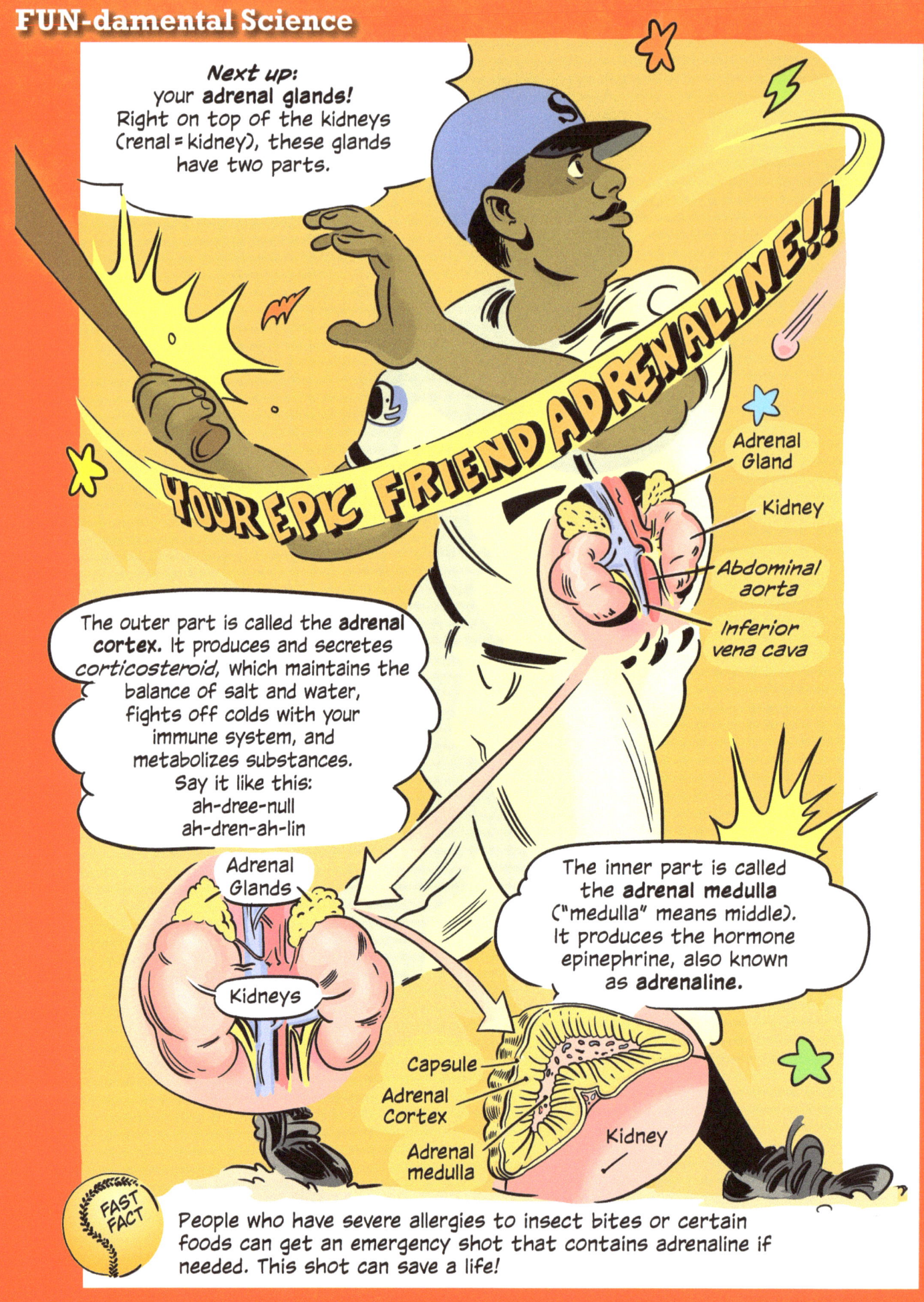

THE ENDOCRINE SYSTEM

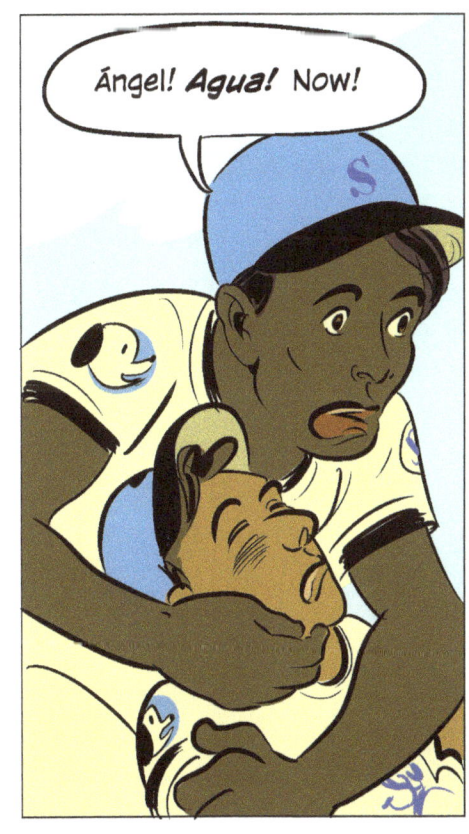

THE ENDOCRINE SYSTEM

THE ENDOCRINE SYSTEM

THE ENDOCRINE SYSTEM

HEY, META, META-SWING!!

Another endocrine gland is the **thyroid**, which regulates the body's **metabolism, growth,** and **development**. Metabolism is a series of chemical reactions in your cells that helps you live. This includes turning food into tissue and energy, as well as releasing that energy and making waste. Say it like this for metabolism: muh-**tab**-o-liz-um metamorphosis: met-uh-more-**fuh**-sis

The thyroid is located in the neck, right at the base of your throat. It produces and secretes thyroid hormones.

Remember: hormones are messages and communicate with your cells. **Shaped like a butterfly,** this gland's hormones are social butterflies when interacting with cells: they target virtually every cell in the body!

¡Vamos, you can do it!

You! Sprint!

You! Slide into home!

FAST FACT

Butterflies are known for *metamorphosis*, their transformation from a caterpillar. The words *metamorphosis* and *metabolism* both have Greek roots that mean "change."

THE ENDOCRINE SYSTEM

SCORE A HOME ostasis RUN!
The Pancreas and Diabetes Defined

The j-shaped **pancreas**, right behind your stomach, is one important player! This gland's role runs deep, in two different body systems.

In the digestive system, the pancreas neutralizes stomach acid and helps break down food into **glucose,** a type of sugar your body uses for energy. The stomach and small intestines absorb this glucose then release it into your bloodstream.

Say it like this:
pancreas: pain-**cree**-us
diabetes: die-uh-**beet**-ee-s
glucose: **glue**-cos
homeostasis: home-ee-oh-**stay**-sis

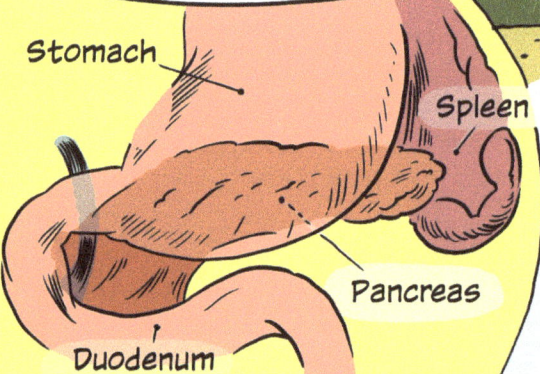

Stomach
Spleen
Pancreas
Duodenum

The pancreas secretes **insulin,*** which "unlocks" your cells so they can use the sugar from your bloodstream, as well as store any excess sugar in your liver.

Diabetes is a condition caused by an imbalance of insulin: **Type I** means your pancreas can't make insulin, and **Type II** means your pancreas is having trouble maintaining insulin levels.

This **imbalance** affects your **blood sugar level**, or the amount of sugar in your bloodstream. Diabetics have to be very aware of body signals and check their insulin levels regularly. Some signals of low blood sugar include hunger, shaking, and lightheadedness.

The goal for your body's endocrine system—and all your systems, really—is **homeostasis**: having the right amount of everything when and where you need it.

Say it like this: "IN-suh-lin."

FAST FACT Type I diabetes is often diagnosed in early childhood, but can be discovered in adults as late as their 30s.

ADVENTURE 8

Learning Calendar

Part 1
Know Your History

Locate Cuba on the world map using a globe, an atlas, an online map, or checkout Google Earth.

Read the comic, **Grand Slam!**
Find it at the beginning of this Adventure Guide!

Gather the adventure equipment from around your house - use the checklist on pages 42 and 43!

Dive into *Know Your History*

Make *Baseball's Secret Language Your Own*

Get to *Know Your Son Cubano*

Uncover *Hidden Message Poems*

Puzzle through *Evolution of the Revolution* Crossword

Recite *Cuban Comprehension*

Part 2
Know Your Endocrine System

Explore *Know Your Endocrine System*

Question *Show Me the Sugar!*

Find *Adrenal Aegis*

Become a *Matchmaker*

ADVENTURE 8

Experience *Endocrine Expertise*

Talk *About a Team*

Part 3
Know Your Appetite

See *Know Your Appetite*, shop for ingredients, and get your kitchen ready!

Prepare *Cuban Style Black Beans* and t

Share your dishes with your family. Discuss *Thoughts for Young Chefs* around the table!

Part 4
Show What You Know!

Finish strong with *In it to Win it!*

Check out *Further Reading* for more opportunities to learn.

Great job on all your hard work!

THE ENDOCRINE SYSTEM

Home Inventory Checklist

Ask your parents to help you find these items around the house. These are some of the tools you will need on your adventure. Check them off the list as you find them!

- ☐ **Two wooden pencils or dowels**
  (Know Your Son Cubano)

- ☐ **Optional: Wonderful Sound Waves Shaker**
  (From Adventure1) (Know Your Son Cubano)

- ☐ **Optional: A metronome, set at 100 beats per minute**
  (Know Your Son Cubano)

- ☐ **Pen**
  (Hidden Message Poems)

- ☐ **Paper**
  (Hidden Message Poems)

- ☐ **Teaspoon**
  (Show Me the Sugar!)

- ☐ **Lots of sugar**
  (Show Me the Sugar!)

- ☐ **Parent or friend**
  (Show Me the Sugar!)

- ☐ **Measuring cup**
  (Show Me the Sugar!)

- ☐ **Clear plastic bag**
  (Show Me the Sugar!)

- ☐ **A carbonated beverage**
  (Adrenal Aegis)

"Be sure to check the items off when you've found them."

- ☐ **Two glasses or cups** (Adrenal Aegis)
- ☐ **Ice cubes** (Adrenal Aegis)
- ☐ **Scissors** (Matchmaker)
- ☐ **Colored markers** (Butterfly Beat)
- ☐ **A popsicle stick** (Butterfly Beat)
- ☐ **Tape** (Butterfly Beat)

"Don't worry if you can't find every single item – just use your imagination to find substitutions!"

THE ENDOCRINE SYSTEM 43

Know Your History

Cuba, 1954

Cuba, Conquered & Traded

After conquering the Cuban natives, the Spanish established Havana. But did you know the city was once (for a short time) in British hands? In 1762, the British captured Havana, held it for less than a year, then traded it back to Spain. The Spanish enslaved and brought hunderds of thousands of African people to Cuba to plant and harvest sugarcane. In 1898, during the Spanish-American War, the United States helped Cuba fight for freedom from Spain. The U.S. controlled the island until the Republic of Cuba was created in 1902, formally gaining its independence.

Havana High Life

After Cuba became independent, Havana developed its unique Cuban feel. The business of growing sugarcane and producing sugar had made the city wealthy. In the 1920s, hotels, restaurants, and nightclubs sprang up. The beautiful beaches, old Spanish buildings, and lively music and dance drew movie stars, like Ava Gardner and Frank Sinatra, and even American mobsters to Havana to vacation, play, and enjoy "The Havana High Life."

ADVENTURE 3

Communist Revolution

Sugar and tourism made some Cubans rich, but not all. There were still a lot of people who lived in poverty. In 1959, communist revolutionaries led by Fidel Castro took over and created the governemnt that rules Cuba today. Cuba changed dramatically after Castro. Private businesses were no longer allowed. The government took control of everything. Professional baseball was banned (although amateur baseball continued to be hugely popular). Castro even banned Monopoly, the board game, because it is a game about private ownership!

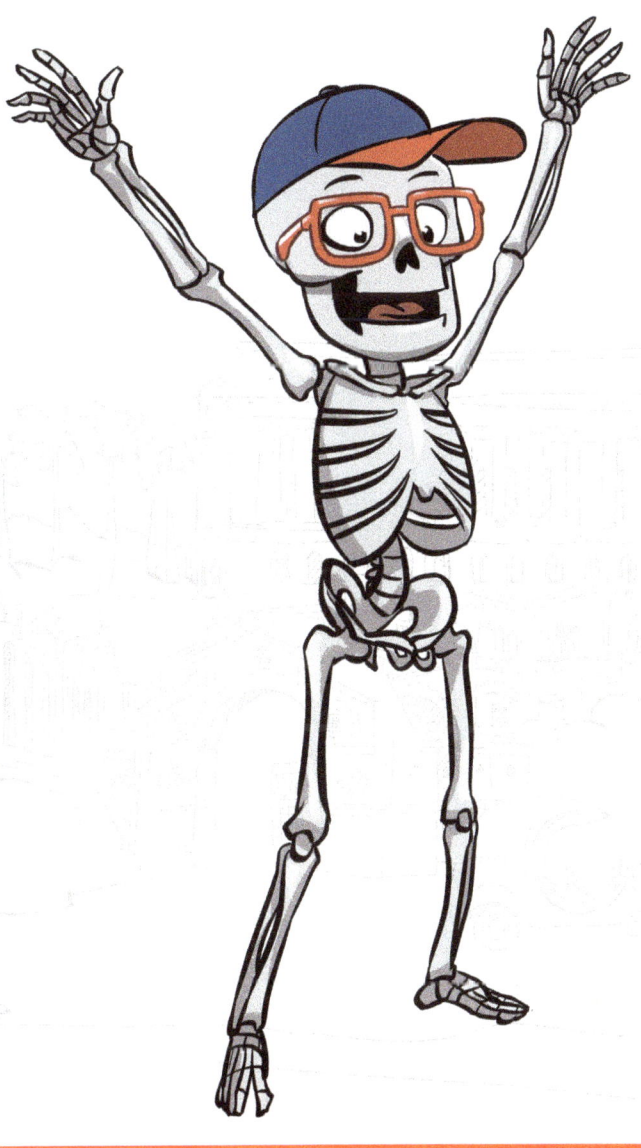

Know Your Cuban Baseball

In Cuba, baseball is not just a game—it's a symbol of resistance and freedom.

Just Try to Ban Baseball

Baseball was introduced to the island in the 1860s by Cuban students returning from colleges in the U.S. and American sailors who sailed into port. Cubans quickly adopted the sport, and leagues started all over the island. But, in 1869, after the first Cuban War of Independence, Spanish rulers banned baseball. They tried to make Spanish-style bullfighting the national sport. Luckily for fans, banning baseball just turned the game into a symbol of resistance against the Spanish.

The Sugar Mill Leagues

In the 1950s, amateur (or non-professional) leagues were the heart and soul of Cuban baseball. One popular form was sugar mill baseball. Each team represented a different sugar mill and the players—sugarcane cutters and factory workers—competed against each other in weekly games. Players in the league may have used baseball as an escape from the harsh working conditions of the mill, but sugar mill baseball produced some of the most talented players in Cuban baseball.

Cuba's Premier Stadium

Estadio del Cerro, currently known as Estadio Latinoamericano, is Cuba's premier baseball stadium. Commonly known to locals as "the Colossus of Cerro," the stadium was built in the municipio (district) of Cerro, which is a small section of Cuba's capital, Havana. The stadium opened in 1946 and has been home to Los Industriales, Cuba's most popular professional baseball team for over 50 years. It can hold 55,000 excited baseball fans!

Make Baseball's Secret Language Your Own

If you've been to a baseball game, you may have noticed the players using gestures with their hands and bodies (right hand to left elbow) to communicate silently with each other. But did you know the real signs are surrounded by a string of false signals (called "dekes" for decoys) to confuse the other team?

In a typical 9-inning game, over 1,000 signals can pass from catcher to pitcher, coach to batter, and between outfielders.

Typically, a set of signs will include a bunch of *"fake signs,"* then an *"indicator"* and a *"hot sign."* The indicator lets the player know that the hot, or important sign, is coming. And the fake signs confuse the other team.

For example, the indicator might be placing the left hand on the right knee, while the hot sign might be to tip your cap. These signs are the only ones that matter, but they are embedded in a series of tummy rubs, arm crossings, nose flicks, and just about anything else the coach can think up.

> Want to communicate with people in secret? Use the idea behind baseball signs to create your own secret language!

Materials:

- **Yourself and one or more partners** - (if you have four or more people, break into teams)

Directions:

1. Find a spot to practice where your team won't be overheard.

2. With your partners, decide what your 'indicator sign' will be. This is a gesture that tells your teammates that an important signal is coming.

3. Create three or four signs for actions. You may want to write them down for later reference. Just remember to hide that piece of paper!

4. Practice a few false signs that don't mean anything, so you can use those as fluidly as the real ones.

5. Once you and your partners have learned the signs, try them in action with the false ones. If you had enough people to form teams, have each team take turns doing the signs while the others try to guess which are real.

Know Your Son Cubano

African Rhythms + Spanish Songs = Son Cubano!

Can you guess why Cuban music has roots in African rhythms? If you thought it's because of the slave trade, you're right. The Spanish enslaved and brought more than a million African people to Cuba, so it's not surprising that Cuban music has been deeply influenced by African traditions.

Son Cubano is a popular Cuban style of music and dance. It combines the sound and feel of Spanish songs with African rhythm and percussion. Not surprisingly, the instruments used to make Son Cubano are also a mix of Spanish and African influences.

There are African-inspired percussion instruments like bongo drums and **clave*** sticks. Also, Spanish-inspired guitars like the tres Cubano, a 3-string guitar. And even a thumb piano called the **marímbula***.

bongo drums

marímbula

clave sticks

*Say them like this:

clave - "CLAW-vay"   marímbula - "mar-im-BUE-la"

The strongest syllable is always shown in CAPITALS and red.

ADVENTURE 8

coloring opportunity

You should join the Bonyfide Band. What instrument do you play? Draw yourself in the band.

THE ENDOCRINE SYSTEM

Make Your Own

Clave Rhythm

Musicians perform music to a steady pulse called the beat. Patterns of accents on the beat are called rhythm. For example, the rhythm of your heart is a loud thump, a soft thump, and a pause.

Clave is a spanish word meaning "code" or "key." It represents the part of the music that holds the rest together, or rhythm. A clave pattern is meant to signal to the other musicians what is happening. These distinctive patterns can range from very complicated to simple, and are what cause music to get stuck in your head.

Materials:

- **Two wooden pencils or dowels**

   (make sure they are not sharp!)

A polyrhythm is a combination of two or more different rhythms. What many musicians know as a "two-three" rhythm is present in a lot of African music. It is also an integral part of Son Cubano music and dance.

52

ADVENTURE 8

Directions:

1. First, count out loud as shown in the chart below, saying "One and two and three and four and" until you feel confident you can keep it going.

2. Hold one of your sticks loosely in one hand, and firmly grip the other.

3. Continuing to count, strike the loosely held stick with the firm one on the beats indicated in the chart. It may take some practice - there's a reason why a band needs someone dedicated to helping keep this pattern going!

1	and	2	and	3	and	4	and
♦			♦				

1	and	2	and	3	and	4	and
		♦					

 indicates a beat

Practice with other patterns - find one that makes you excited!
See if you can find music that contains a clave pattern - *it can be found in all sorts of music these days.*

THE ENDOCRINE SYSTEM

Know Your Cuban Poetry

Take a revolutionary spirit and a love of words, mix together and who do you get? **José Martí**—Cuban freedom fighter and poet! Born in 1853, he lived at a time when Cuba was trying to break away from Spain. He fought, wrote, and died for Cuban independence.

Writing & Fighting for Freedom

Martí disliked the fact that Spain ruled his homeland. When he was just 16, he was sentenced to hard labor (breaking big rocks into small rocks) as punishment for resisting the Spanish. He was then thrown out of Cuba. He eventually came to the United States, where he became famous for his poems about Cuba, justice, and freedom. He also made plans to return to Cuba and fight the Spanish. On April 11, 1895, he and his supporters arrived on the island and the battle began. Martí did not survive but his poems live on.

Hidden Meanings

One of Martí's most famous poems is called *Los Zapaticos de Rosa*, or *The Pink Shoes*. Many Cuban children learn to memorize and recite it in school. Like many of his poems, it is more than just a story.

In Martí's poem, a young girl named Pilar gets dressed up in a feathered hat and fancy pink shoes, and goes with her mother to the beach. Her mother warns her not to get her shoes dirty, but when Pilar comes back from playing, she has no shoes. "Where are your shoes?" cries her mother. Just then a poor woman comes to explain. Upon meeting the poor woman's daughter, who is sick and has no shoes at all, Pilar gave the child her beautiful pink shoes.

Based on the summary, can you find the hidden meaning in this poem? What do you think it is?

_____
_____
_____
_____
_____

What might Marti be saying about society?

_____
_____
_____
_____
_____
_____

Hidden Message Poems

Why do poems often seem mysterious and puzzling? Like José Martí, poets use words to share feelings and ideas. This can make poems feel like they have hidden messages or meanings.

It's fun to use poetry to share a hidden message—perhaps with friends who know you really well, so that only they will get your meaning. One way to place a message in a poem is to place one word from your message into each line.

Materials:
- **Lined or Blank Paper**
- **Pen or Pencil**

Directions:

1. In order to use your code, the person you are sending it to needs to know how to read it. It can be as simple as the first or last word in each line or something more complicated. For example, in the poem titled "In The Year 1954," you would let your friend know to read the 1st word in the first line, 9th word in the second line, and so on. Let your friend know the code in advance, so they can be ready whenever you have a secret message to send.

2. Come up with the message you want to share and write it down - one word on each line Remember, you'll be coming up with a line of poetry for each word in your message, so you may want to keep your message short.

3. Decide what you want your poem to be about. How do you want it to connect with the hidden message? Maybe your poem expands your message, has the opposite meaning or is entirely unrelated. It's up to you.

4. Now you can begin to fill your poem in. Remember what your code is, and make sure the words you chose are in the right spots. If you have trouble, consider changing the words in your hidden message to make it easier.

5. When you are finished, you should have a normal-looking poem, with a secret message hidden to be read from top to bottom.

When you are finished, you should have a normal looking poem, with a secret message hidden, line to line, from top to bottom.

Here's an example.

When we **meet**
You will see the real **me**
Not hidden like a blade of grass in the **park**
But blazing like the sun at **noon**

THE ENDOCRINE SYSTEM

Know Your Cuban Language

While the official language of Cuba is Spanish, the story of language in Cuba is wide and diverse.

Spanish was first brought to Cuba by colonizers in 1492, but it hasn't stayed the same since. Over time it morphed into its own variant, **español cubano.** Español cubano, or Cuban Spanish, has a record of all the historic influences on the island contained within it including words originating from the Canary Islands, West Africa, France, English and the indegenous people.

Cuba uses the word *guagua* for a bus, in Spain it would be autobus.

In Argentinian Spanish, though, *guagua* means baby!

guagua   *guagua*

ADVENTURE 8

Spanish is also not the only language spoken in Cuba.

The second most popular language is Haitian Creole, a mixture of French vocabulary and West African grammar that originated among the slaves the French brought to their colony in Haiti. When those slaves revolted and formed their own government the language stayed, and immigrants from Haiti brought it to Cuba where it's a recognized language.

In 2014, the Cuban government started working on a project with Haiti to promote reading in Creole to help protect the dialect and bring the two island nations closer together.

Other languages spoken in Cuba include Galician, Corsican, and Lucumi!

THE ENDOCRINE SYSTEM

Know More Spanish

Even though the Spanish Empire no longer rules over Cuba, the language still remains an important part of Cuban culture. Check out a few Spanish phrases below! They might just come in handy one day.

"Hello, my name is Dr. Bonyfide."

Spanish Phrase: "Hola, mi nombre es Dr. Bonyfide."

Say it like this: "Oh-la mee gnome-bray-eh-s"

English Phrase

"Good Morning"

"Good Afternoon"

"Good Evening"

"How are you"

"Thank you"

"How much does this cost"

"Where are the bathrooms"

"Do you speak English"

ADVENTURE 8

Spanish Phrase	Say It Like This
"Buenos días"	"Bu-when-ohsd-ee-us"
"Buenas tardes"	"Bu-when-ohs tar-des"
"Buenos noches"	"Bu-when-ohs no-ch-es"
"Cómo estás"	"Koh-mo es-tahs"
"Gracias"	"Gra-see-ahs"
"Cuánto cuesta este"	"Qu-ahn-tow quest-ah es-teh"
"Donde esta los baños."	"Don-day es-tah lo-s bah-nyo-s."
"Habla Inglés"	"Ah-blah-Ing-l-es"

Evolution of the Revolution

ADVENTURE 8

Across:

1. Name of a Cuban freedom fighter and poet

4. The name of a Cuban dish combining rice and black beans

5. A sport that symbolizes resistance and freedom in Cuba

7. Cuban food is also known as _____ food

8. The American mainland state located closest to Cuba

9. A steady pulse musicians play to

Down:

2. A crop in Cuba that made some very rich

3. Means, "Made up of words that have hidden messages or meanings"

5. Means, "Welcome" in Spanish

6. A culture that influenced percussion instruments, like bongo drums and clave sticks

You can check your answers using the key on page 118.

Cuban Comprehension

Good work, Adventurers!

Now that you have read some things about Cuba, let's review what you have learned!

Try to fill in the blanks.

Havana was established by the _ _ _ _ _ _ _ , though it was briefly captured by the _ _ _ _ _ _ _ _ . The Spanish enslaved and brought African people to Cuba to harvest _ _ _ _ _ _ _ _ _ _ . In _ _ _ _ the Republic of Cuba was created, gaining _ _ _ _ _ _ _ _ _ _ for the Cubans. After becoming independent, the _ _ _ _ _ industry made many wealthy, but many still lived in _ _ _ _ _ _ .

In _ _ _ _ the revolutionaries under _ _ _ _ _ _ _ _ _ _ _ _ _ _ took over, trying to create a communist society. The government took control of everything, even banning professional _ _ _ _ _ _ _ _ and the board game _ _ _ _ _ _ _ _ .

You can check your answers using the key on page 119.

When a million African slaves were brought to Cuba, they brought with them their music and formed the style __ __ __ __ __ __ __ __. It uses a mix of __ __ __ __ __ __ __ songs with __ __ __ __ __ __ __ rhythm and percussion. The Cuban instrument of the __ __ __ __ __ is used in these rhythms. Cuban students brought back the sport of __ __ __ __ __ __ __ __ in the 1860's from trips to the U.S. and it became quickly popular. When an attempt to gain independence failed in 1869, the Spanish banned it, trying to replace it with __ __ __ __ __ __ __ __ __ __ __!

This turned it into a symbol of __ __ __ __ __ __ __ __ __ __.

THE ENDOCRINE SYSTEM 65

Know Your Endocrine System

The teamwork performed by glands in your endocrine system regulates, controls, and coordinates many bodily functions. It is in charge of communication between cells, tissues, and organs, using signals called hormones. Hormones are like chemical sparks.

Here is Your Endocrine Team Lineup!

Hypothalamus
General Manager: Processes information from the brain and directs messaging between hormones and cells.

Pituitary
Assistant Manager: Stores and releases hormones made by the hypothalamus and regulates other glands.

Thyroid
Regulates metabolism, growth, and development. Thyroid hormones are like social butterflies, trying to communicate with EVERY cell!

Adrenal
Influences metabolism and response to stress. Two parts, cortex and medulla, produce corticosteroid and adrenaline hormones.

They excite and ignite cells into action. Your endocrine glands make hormones and secrete them directly into your bloodstream, to communicate with special "target" cells.

Like a key fits a lock, cells with receptors respond only to specific hormones. These hormones and cells help you absorb energy from the food you eat. They also help you grow, and they are sometimes even responsible for changes in your moods.

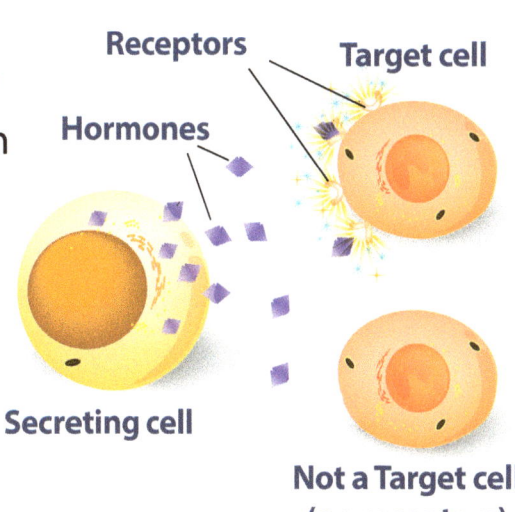

Just like a baseball team!

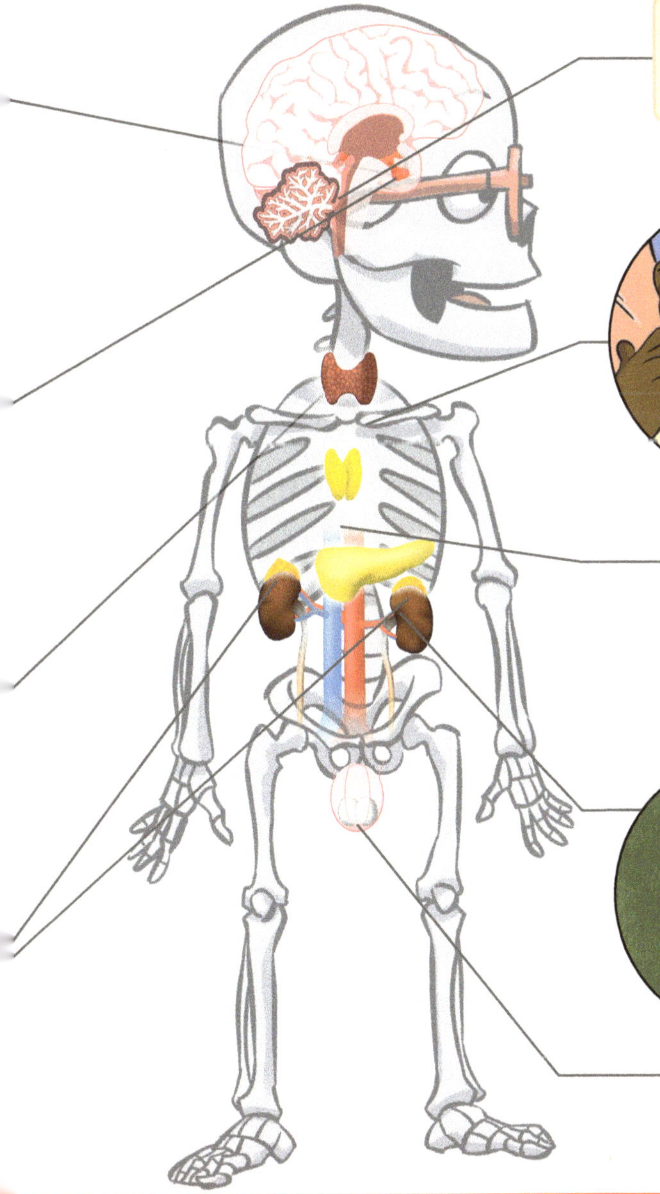

Pineal
Helps regulate sleep patterns by secreting the hormone melatonin.

Parathyroid
Regulates the balance of calcium in your body and tells bones to release some, if needed.

Thymus
Only active until puberty, produces and secretes hormone thymosin. Also stimulates T cells for the immune system.

Pancreas
Switch Hitter: Regulates levels of sugar in the blood and helps break down foods in the digestive system.

Reproductive
Also called gonads, makes and secretes hormones that control mood. Testes for boys and ovaries for girls.

THE ENDOCRINE SYSTEM

On The Bench

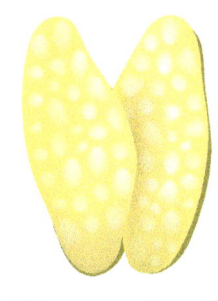
Thymus gland

Remember the **thymus** from the comic? Even though it's not a main gland in the endocrine system, and is only active until puberty, the thymus is a secondary gland and plays an important role in your immune system.

Located in the chest behind the sternum and between the lungs, the thymus produces the hormone thymosin, which stimulates the development of disease fighting T cells. T Cells work with the immune system to fight infections and prevent auto-immunity, when the body incorrectly turns against itself.

Also sometimes overlooked on your endocrine team bench is the **parathyroid**, which regulates your body's calcium levels. The four parathyroid glands are found on the back of your butterfly-shaped thyroid, kind of like the eyespots on a butterfly's wing.

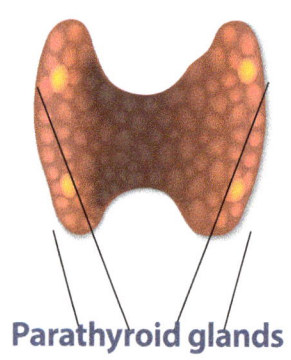
Parathyroid glands

Your parathyroid is on the job all the time, making sure there is the right amount of calcium in your blood. Healthy calcium levels help your muscles and nerves function at their best. If you don't have enough calcium, your parathyroid tells your bones, which store lots of calcium, to release some into your bloodstream. Parathyroid hormones also help Vitamin D become active in your body, leading to stronger bones.

Zoned Out

Have you ever zoned out and missed out on something important? It might be because you did not get enough sleep. Do you remember the last time you tried to stay up all night? Were you able to stay up, or did you close your eyes and doze right off to sleep? If you did the latter, **melatonin*** played a role in making that happen.

Pineal gland

The small, pinecone-shaped pineal* gland, located exactly in the middle of your brain, is responsible for secreting the hormone melatonin. This hormone helps regulate when you fall asleep at night as well as when you wake up in the morning. It also affects the hormones related to the reproductive glands.

---

*Say them like this:

pineal - "pin-e-ul"

melatonin - "mel-uh-toe-nin"

The strongest syllable is always shown in **CAPITALS** and red.

THE ENDOCRINE SYSTEM

Ch-Ch-Changes: Never a Dull Adolescence

Hormones are some of the heavy hitters in your endocrine system. They help you change into a teenager—and eventually an adult. At some point, usually between age 11 and 14, pimples might start popping up on your face—sometimes even overnight! Maybe moodiness has become your middle name.

This time of change is called **puberty**.* Your body is starting to turn into an adult body, thanks to reproductive glands—called **gonads**.* In girls, the gonads include two ovaries—attached to the uterus in the lower abdomen—and in boys, two **testes**,* in the scrotum.

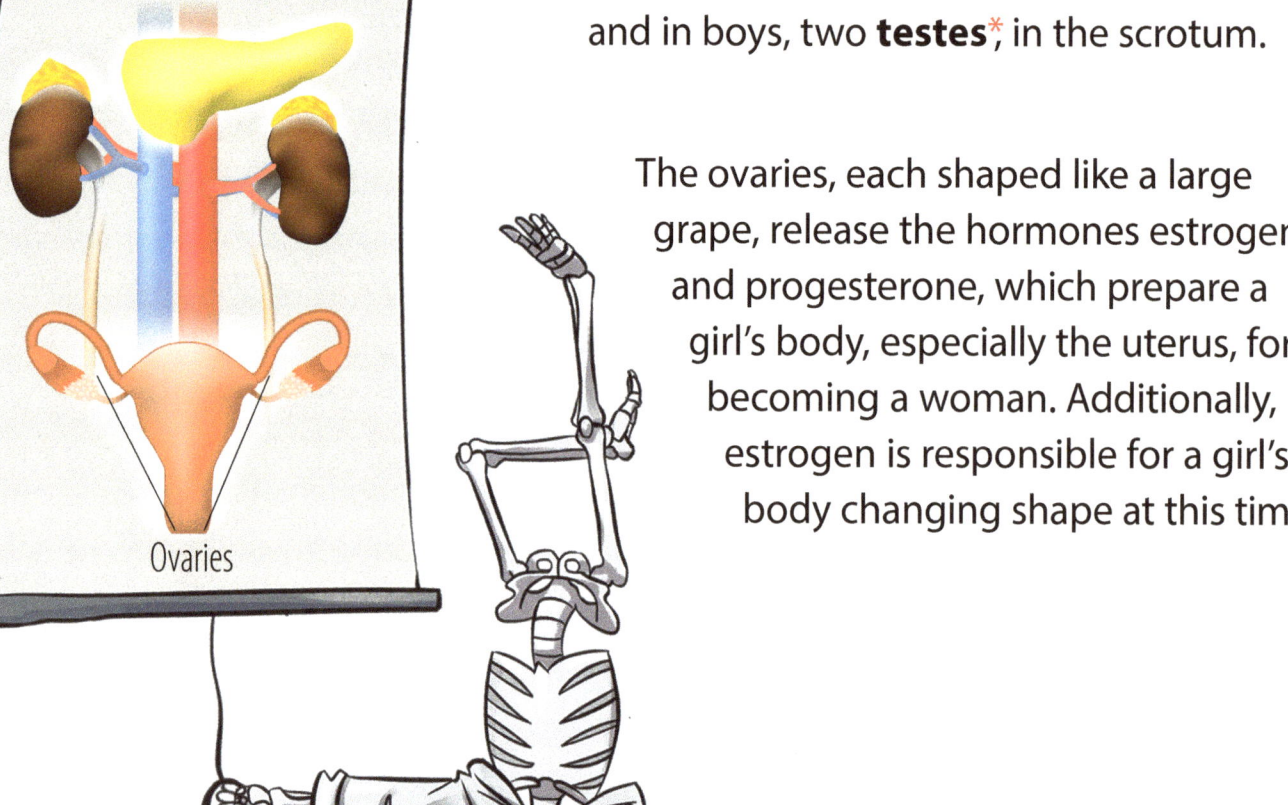

Ovaries

The ovaries, each shaped like a large grape, release the hormones estrogen and progesterone, which prepare a girl's body, especially the uterus, for becoming a woman. Additionally, estrogen is responsible for a girl's body changing shape at this time.

ADVENTURE 8

The testes release a family of hormones called androgens. One of the most important is testosterone.* For boys, this hormone is responsible for the new hair that sprouts unexpectedly, or a new, sometimes creaky, deeper voice.

> For boys, this hormone is responsible for new hair that sprouts unexpectedly, or a new, sometimes creaky, deeper voice

Although the **gonads*** produce and secrete hormones, they are under control of the hypothalamus-pituitary complex. In order for the gonads to know when to start producing and secreting hormones, the hypothalamus sends signals to the pituitary to send instructions in the form of hormones. After receiving the signals and instructions, gonads know exactly what to do!

---

*Say them like this:

puberty - "**pyoo-ber-tee**" **gonads** - "**go-nads**"

testes - "**tes-tees**"

The strongest syllable is always shown in **CAPITALS** and red.

THE ENDOCRINE SYSTEM

Know Your Pancreas

Finding the Sweet Spot

One of your largest glands, your six-inch-long pancreas, lies deep within your abdomen between your stomach and your spine. This switch hitter bats for both your digestive and endocrine systems to regulate the level of sugar in your blood.

A small section of cells within the pancreas create specific hormones, **insulin*** and **glucagon.***

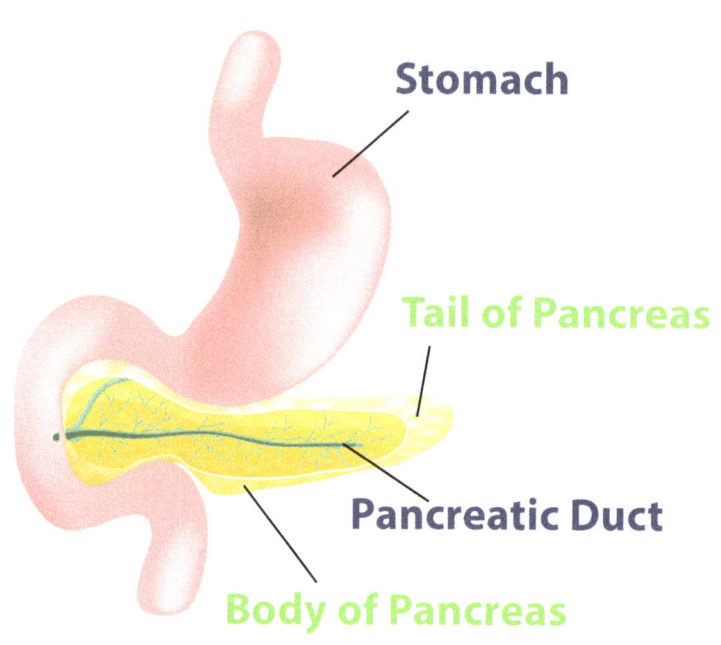

Insulin and glucagon work in opposite ways. When your blood sugar levels are too high, insulin tells your cells to absorb and use glucose. When your blood sugar levels are too low, glucagon tells your cells to release glucose because your body needs this energy.

But sometimes the pancreas doesn't do its job properly, and has trouble maintaining insulin levels. When it does not produce enough insulin—sometimes none at all—the condition is called type 1 diabetes and it is more common in kids than adults.

> *Say it like this:
>
> **insulin** - "in-suh-lin" **glucagon** - "glu-cuh-gon"
>
> The strongest syllable is always shown in CAPITALS and red.

72

ADVENTURE 8

coloring opportunity

THE ENDOCRINE SYSTEM 73

Show Me the Sugar!

Most people think of sugar as candy.

However, people of all ages are consuming more sugar in the form of "added sugars" in pastries, snacks, sodas, juices, sports drinks, and other beverages.

Have you ever heard of these? They are examples of added sugars listed as an ingredient in your favorite snack or drink…

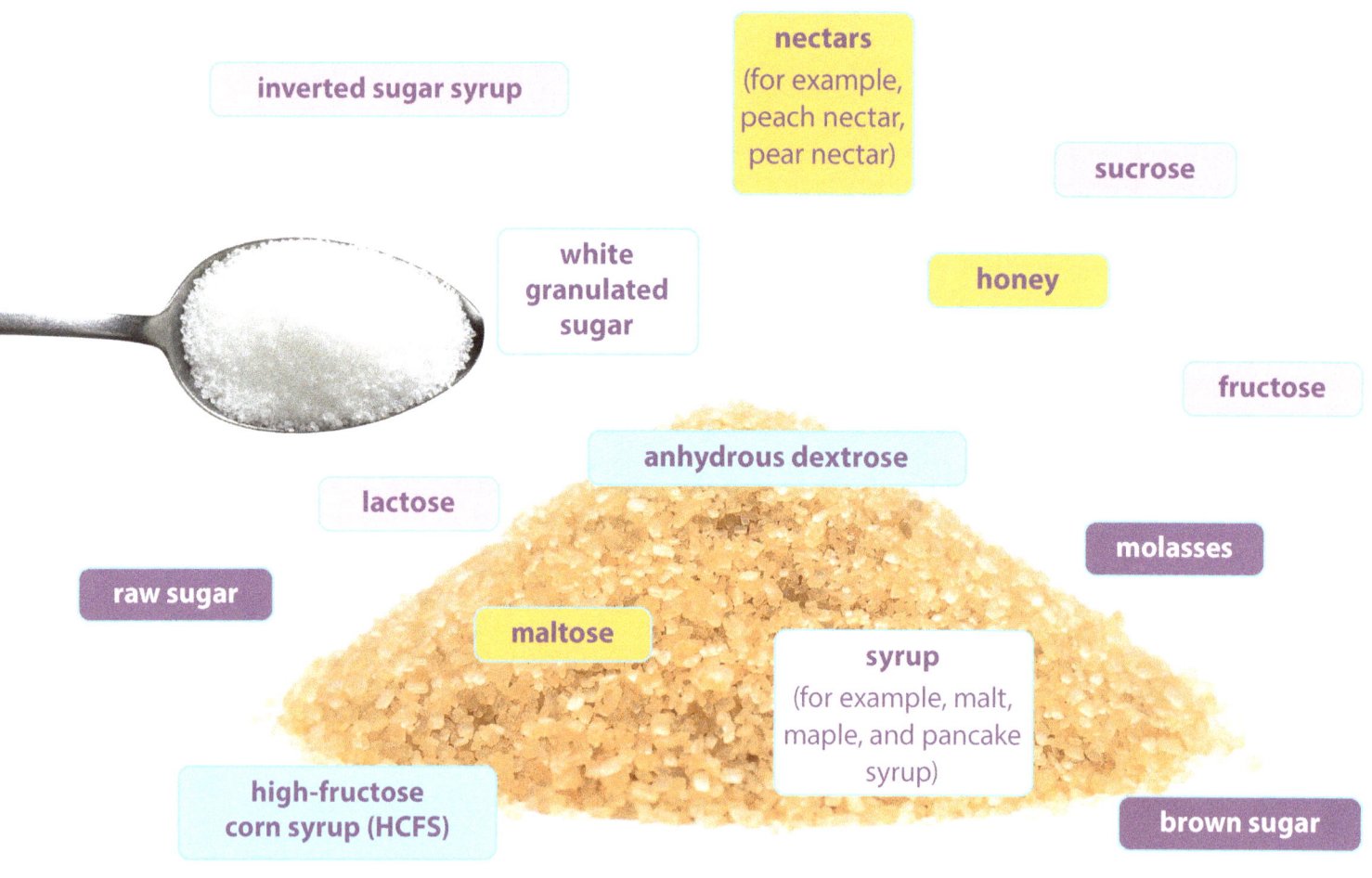

- inverted sugar syrup
- nectars (for example, peach nectar, pear nectar)
- sucrose
- white granulated sugar
- honey
- fructose
- anhydrous dextrose
- lactose
- molasses
- raw sugar
- maltose
- syrup (for example, malt, maple, and pancake syrup)
- high-fructose corn syrup (HCFS)
- brown sugar

What can happen if your eating habits over time become highly concentrated in added sugars? Everyone responds differently to excessive amounts of sugar but some people develop greater risk for diseases, including type 1 **diabetes.**[*]

Your body is constantly seeking balance in all things. Specifically, your endocrine system is working hard to balance blood sugar levels.

We may not realize how much sugar we consume every day. Just because we don't see the sugar, doesn't mean it's not there.

4 grams of sugar ~ 1 teaspoon of sugar
The recommended daily amount of sugar for a young person is bettween 6 and 8 teaspoons, or 24 to 32 grams.

Type 1 means your pancreas can't make insulin, and type 2 means your pancreas is having trouble maintaining insulin levels.

THE ENDOCRINE SYSTEM

Show Me the Sugar

Let's pretend, in one day, your parent or friend consumed the following items:

Materials:

- **Teaspoon**
- **Lots of sugar**
- **Measuring Cup**
- **Clear Plastic Bag**
- **Parent or Friend**

Directions:

1. Prepare the following amounts of sugar and pour into a plastic bag.
    a. 8 teaspoons ~ Sports drink
    b. 8 teaspoons ~ Muffin
    c. 4 teaspoons ~ Apple sauce

2. Tie the plastic bag in a knot, hold it up and show your parent or friend. Ask, "How much sugar do you think this is?" and "Would you eat this much sugar?"

3. Open the bag, and pour the sugar into the measuring cup.

4. Show them the sugar and say, "Guess what! You just had a muffin, an apple sauce and a sports drink!"

Discuss:
- The total amount is almost half a cup of sugar. Is that a lot?
- Was your friend shocked by the amount of sugar consumed today?
- The World Health Organization recommends that kids consume 6–8 teaspoons of added sugar per day.
- What would you not eat to stay below the 6–8 teaspoons in this activity?
- What would you choose to eat the rest of the day to remain below 6–8 teaspoons?

Tips for Success:

Sip smarter!
Dilute your beverages or choose water instead!

Be an inventor!
Create new combinations of whole grains, fruits, and nuts. Package them in fun ways for on-the-go yum!

Sugar Check
How many food items in your kitchen have added sugars? Check the nutrition labels.

Explore More

In the table below, record your daily intake of sugar for five days. Sugar information is on the nutrition label.
Remember, 4 grams of sugar ≈ 1 teaspoon of sugar.

	Track your daily grams of sugar					
	16g	24g	32g	40g	48g	56g
Monday						
Tuesday						
Wednesday						
Thursday						
Friday						

THE ENDOCRINE SYSTEM

This Adventure's Touchstone: where anatomy, physiology, and psychology all come together.

Sign of the Times: Your Hormone Cascade

When your body senses stress or some possible threat, it protects you by preparing you to handle the situation. Coordinated by your endocrine system, your **"fight or flight"** or **"freeze"** response is important and necessary for survival.

Everyone's endorine system operates a little differently. When your brain senses danger or excitement, the stress horomone adrenaline releases in a rush, like a waterfall!

You might get a fight or flight response:

- Rapid heart rate
- Muscles tense up
- Suddenly feeling antsy and agitated

You might get a freeze response:

- Not being able to think
- Feeling super tired
- Not wanting to move

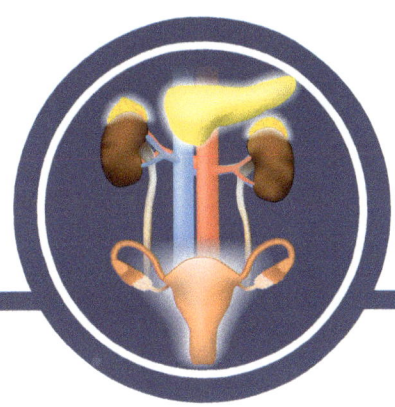

This might look like laziness, but actually your body could be overwhelmed by stress. When you know something important is happening in your life that might cause feelings of nervousness or anxiety, it helps to pay attention to your body signs. Do your hands feel clammy? Can you tell your heart is beating faster? Is it hard to focus?

Try these tips:

- Take 3 deep breaths
- Take a walk
- Read a book
- Have a snack
- Do yoga

THE ENDOCRINE SYSTEM

Adrenal Aegis

Your body has two adrenal glands! the cortex and the medulla

The **adrenal cortex** controls the release of hormones that regulate your body's breakdown of fats, proteins and carbohydrates, and also helps keep your blood pressure under control. It also produces hormones that regulate your immune system. You would not last long without your adrenal cortex helping out every day.

The other gland is the **adrenal medulla**. You need the cortex every day, but the medulla serves a different purpose: emergencies. When your nervous system tells the body you are stressed, the adrenal medulla kicks into action.

This is an important part of the fight-or-flight response, which helps give you the extra energy to get through dangerous situations.

In the long term increasing your heart rate, rushing blood to your muscles and blood and spiking your blood sugar level wouldn't be good for you but it can help you get out of a sticky situation.

Materials:

- **A carbonated beverage**
- **Two glasses or cups**
- **Ice cubes**
- **Some towels just in case**

Directions:

1. Set up your two cups next to each other, one with ice in it and the other without.

2. Pour the carbonated beverage into the cup without the ice. There should be some amount of fizz.

3. Next, pour the carbonated beverage into the cup with ice. Observe what happens. Just like your adrenal medulla helps to kick off a reaction, the ice helps make much more fizz.

> Ice only causes a small reaction, but you'd need to be very careful if you used mentos. Talk about too much stimulation!

Notice your Adrenal-ine

In Adventure 3, we learned about the circulatory system and how our body reacts to stress. One common result of stress was an increased heart rate or pulse.

This happens because your body begins pumping more blood toward your muscles to help you fight dangers.

Next time you feel scared or stressed, notice how adrenaline hormones are affecting your heartbeat. You can do this by placing one hand over your heart and listening for a 'lub-dub' sound. Do you notice your heart beating faster or slower? Write down your adrenaline observations.

When you are ready to slow down your stress response and slow down your heartbeat, you can try a few relaxation activities - try deep breathing, yoga moves, coloring, or some meditation. Take ten minutes to close your eyes and deeply breathe, slowly stretch out your entire body, or meditate by taking a comfortable seated position and refocusing your thoughts to just the present moment.

coloring opportunity

THE ENDOCRINE SYSTEM 83

Butterfly Beat

A thyroid can have disorders or diseases when it is not working properly.

When a thyroid is overactive, it releases too much hormones into the blood - the name of this is called **hyperthyroidism**.* The opposite problem a thyroid could have is making too few hormones for the bloodstream- this is known as **hypothyroidism**.*

In Know Your Endocrine System, you may have noticed that the thyroid kind of looks like a butterfly!

Have you noticed the thyroid looks like a butterfly?

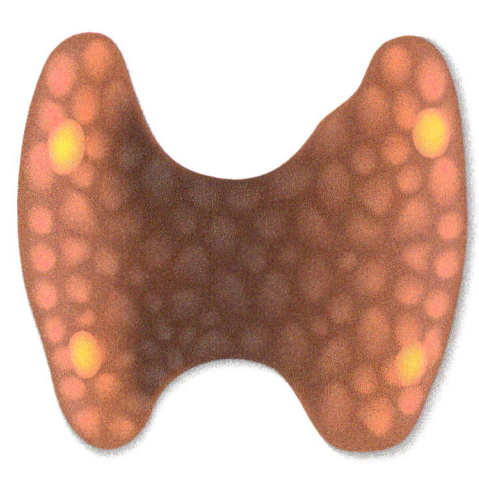

*Say them like this:

hyperthyroidism - "hi-per-TH-EYE-roy-diz-em"

hypothyroidism - "hi-po-TH-EYE-roy-diz-em"

The strongest syllable is always shown in **CAPITALS** and **red**.

ADVENTURE 8

Adventurer, try making a butterfly below!

Can you make the wings go faster and slower?

Materials:

- **Butterfly printout** (page 86)
- **Scissors**
- **Optional: Colored markers**
- **A popsicle stick**
- **Tape**

Directions:

1. Cut out the butterfly design from the printout. Decorating the butterfly with your markers before step 2 is optional!

2. Follow the fold lines on your butterfly to create a square section on the back - this will be used to attach your popsicle stick control.

3. Tape the popsicle stick to the square section on the back.

4. Cut out a long rectangle strip. Find the middle of the strip and tape it to the opposite end of the stick. Once secure, take each end of the paper and tape it to either butterfly wing.

5. Using your fingers to gently pull the paper strip on the back side of the butterfly close to the stick. This will move the butterfly's wings.

When the wings beat slowly, this is similar to hypothyroidism. Beating quickly is like hyperthyroidism.

THE ENDOCRINE SYSTEM

Butterfly Beat Printout

Front

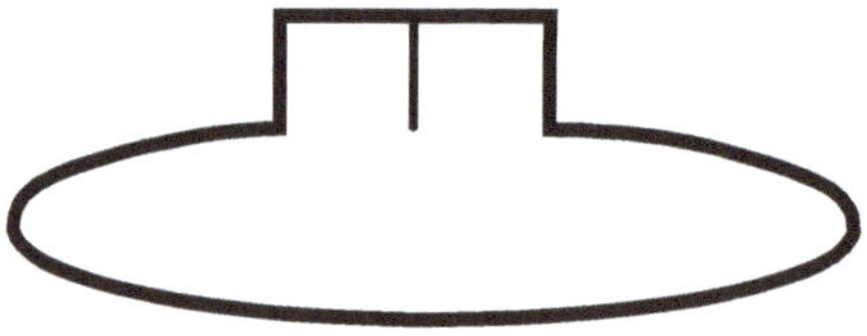

Attach to Back of Body
The two squares will need to be cut and folded opposite directions to attach

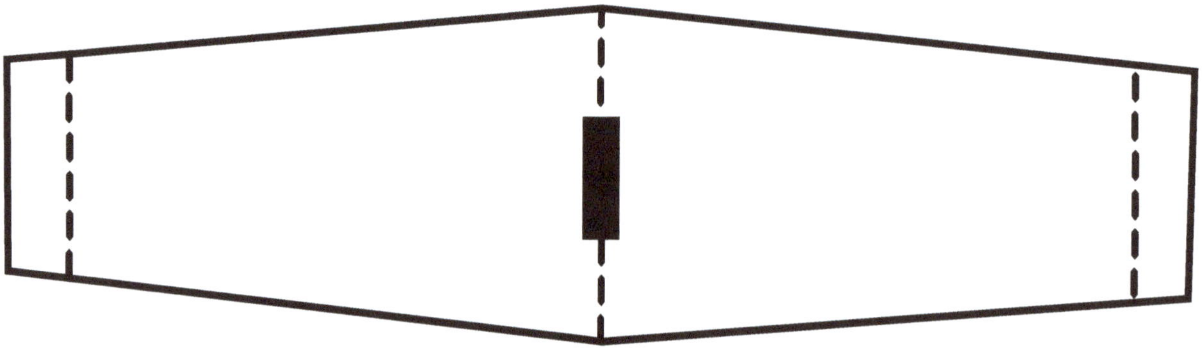

Back.
Fold at dotted lines. Cut at solid lines (inlcuding square in center)

ADVENTURE 8

Matchmaker Printout

Pancreas	Regulates sugar levels in the blood	Adrenal	Influences your metabolism and stress response	Gonads	
Pineal	Secretes the hormone melatonin	Pituitary	Works with the Hypothalamus to regulate other glands	Thymus	
Parathyroid	Regulates calcium in your body	Thyroid	Regulates your metabolism, growth, and development	Hypothalamus	
Gonads	Makes and secretes hormones that control mood	Pancreas		Adrenal	
Thymus	Produces and secretes the hormone thymosin	Pineal		Pituitary	
Hypothalamus	The "General Manager" of the Endocrine system	Parathyroid	P.A.T	Thyroid	

THE ENDOCRINE SYSTEM

Matchmaker

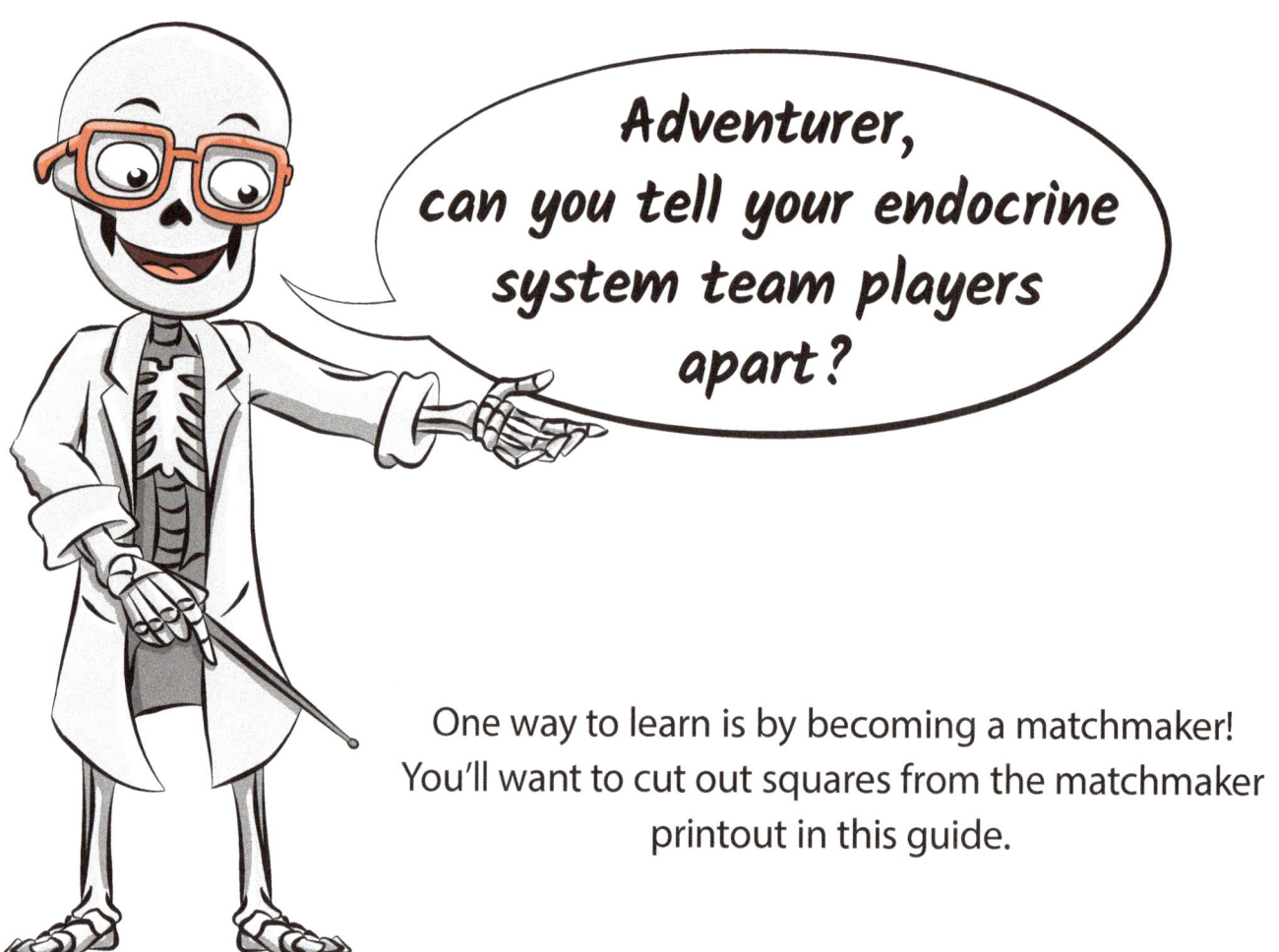

"Adventurer, can you tell your endocrine system team players apart?"

One way to learn is by becoming a matchmaker! You'll want to cut out squares from the matchmaker printout in this guide.

Can you add a couple more pairs to the squares using acronyms and abbreviations? You may have heard someone you know use "BRB" instead of saying "Be right back". This is an example of an acronym, which is made from the first letters of multiple words. Abbreviations are used to shorten long words, you have probably seen one at your Doctor's office - written as "Dr."

Try coming up with your own acronym or abbreviation for glands in the endocrine system - Hypothalamus, Pituitary, Thyroid, Adrenal, Pineal, Parathyroid, Thymus, Pancreas, and Gonads. Then add them to your squares.

Grab a partner and pair up each gland with their function or acronym.

ADVENTURE 8

Materials:

- **Matchmaker printout**
- **Scissors**
- **A partner**

Directions:

1. Print and cut out the squares on your Matchmaker sheet. In the blank boxes, add matching pairs of your own abbreviations. We have filled out Parathyroid for you as an example!

2. Shuffle the cards and lay them on the table, face down in rows.

3. Place all the squares face down and align them in even rows or columns.

4. Grab a partner and decide who will go first. On each turn, a player turns over any two cards (one at a time).

5. If the cards match (for instance, player one turns over Pineal and Secretes the hormone melatonin) the player keeps the cards and gets another turn. If the cards do not match, those cards are turned face down again (in the same position) and it becomes the next player's turn.

6. The trick is to remember which cards are where. The person with the most pairs at the end of the game wins!

You can check your answers using the key on page 120.

THE ENDOCRINE SYSTEM

Chemical Constitution

Oh no! All of Dr. B's glands have gotten garbled! Help him put them back in the right spots!

Match these glands with their proper location.

PINEAL   PANCREAS

REPRODUCTIVE   THYROID

HYPOTHALAMUS   PARATHYROID

THYMUS   PITUITARY

                    ADRENAL

You can check your answers using the key on page 121

ADVENTURE 8

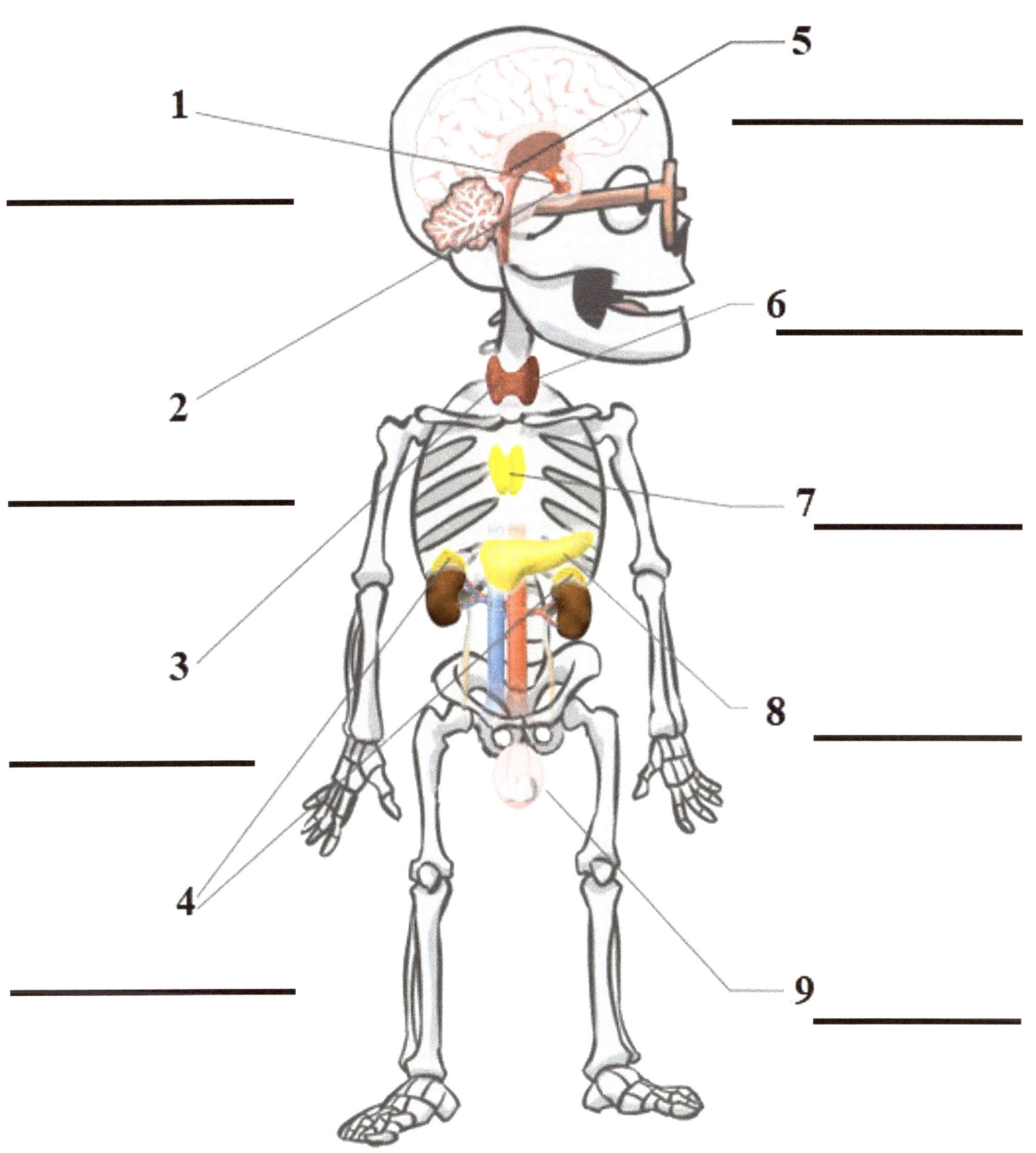

THE ENDOCRINE SYSTEM

Endocrine Expertise

Know Your Endocrine System

HYPOTHALAMUS   MEDULLA

PITUITARY   PINEAL

THYROID   PARATHYROID

ADRENAL   THYMUS

CORTEX   PANCREAS

                    REPRODUCTIVE

                    SUGAR

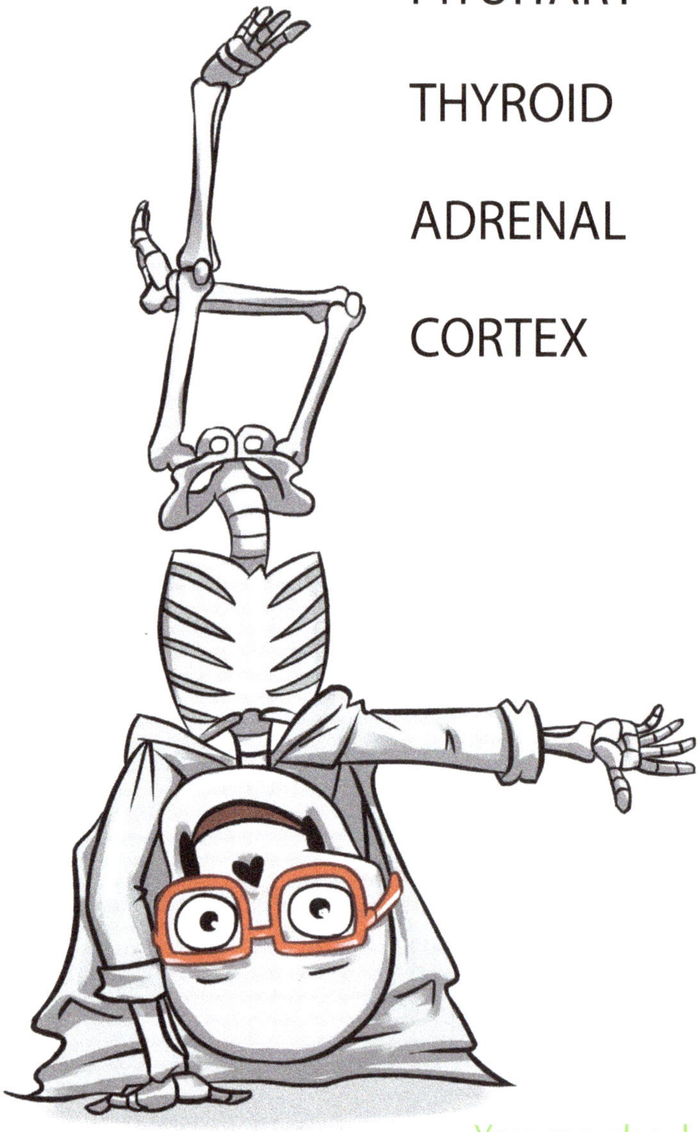

You can check your answers using the key on page 121.

```
W S P R M S H I A F B Z M D G A P C
F T S A E E M Y K E Q P S P C D J G
T P P Q R P D K P R Z B P I X R L C
Y H I N R A R U F O U U H N G E C J
M E Y T H B T O L Z T V J E X N F E
T E B R U W S H D L Z H X A S A M U
J Y Q W O I V U Y U A L A L O L B U
T S J A M I T F G R C M M L G V L T
L E O T E T D A B A O T G J A N J B
T H Y M U S B W R V R I I L J M I Y
C O R T E X W N U Y Y W D V F W U F
W X U P P A N C R E A S Y U E P S S
```

THE ENDOCRINE SYSTEM

Talk About a Team

Good work, Adventurers! Now that you know the endocrine system, let's review what you've learned!

Try to fill in the blanks.

Your e _ _ _ _ _ _ _ _ system is responsible for regulating, controling and

c _ _ _ _ _ _ _ _ _ _ _ many bodily functions. It uses h _ _ _ _ _ _

_ _ which act like chemical sparks to communicate between cells, tissues, and

o _ _ _ _ _.

Ready to verify what you have learned?
See the answer key on page 120

Batting up your endocrine system for the big game (a very important one!) is your h _ _ _ _ _ _ _ _ _ _ _ _, also known as the "General Manager." The p _ _ _ _ _ _ _ _ gland works with the hypothalamus to regulate other glands. Other gland team players are the t _ _ _ _ _ _, a _ _ _ _ _ _ _, p _ _ _ _ _ _, p _ _ _ _ _ _ _ _ _ _ _ _, t _ _ _ _ _ _, p _ _ _ _ _ _ _, and r _ _ _ _ _ _ _ _ _ _ _. The reproductive gland can be called g _ _ _ _ _, one key difference in genders is that boys have t _ _ _ _ _ and girls have o _ _ _ _ _ _.

Know Your Appetite

Experience Cuban Foods

Just like many other aspects of Cuban culture, Cuban cuisine has influences from a wide variety of other cultures, most noticeably, those of Spain and Africa. Because Cuban food is such a blend of outside influences, it has come to be called "comida criolla," which means "creole food," in Spanish. (Creole refers to the food's roots in both Europe and Africa.)

A combination of rice and black beans, a dish called "congrí" is a perfect example of the blend of Spanish and African cultures.

Root vegetables are an important ingredient commonly used in Cuban cooking. Some of the most common root vegetables are yucca, malanga, and boniato. These probably don't sound familiar to you, but they are similar to other root veggies such as sweet potatoes and turnips–and are just as popular in Cuba.

Pinky's Hint:

Read through the entire recipe before beginning to prepare food. This way, you'll know what equipment and ingredients are needed, and you'll be familiar with the steps involved.

 Whenever you see the chef's hat icon, it means **you'll need an adult's help**.

Buen Provecho!

That means

"Enjoy the food!"

in Spanish.

*Say it like this:

"bwen pro-VE-cho"

The strongest syllable is always shown in **CAPITALS** and **red**.

Recipes and food knowledge provided by Chef Polly Legendre of La Gourmande Catering.

Cuban Style Black Beans

Prep time 10 minutes | Cook time 10 minutes

Black beans are a basic ingredient in Cuban cuisine and often featured in daily meals. When the black beans are prepared for "congrí," the white rice takes on the inky color of the black beans, representing the fusion of cultures from Spain and Africa.

When preparing beans without rice, the "sofrito" gives the beans their flavor. Sofrito is a fried blend of onions, peppers, garlic, and oregano and serves as a base for many dishes. It is also an important part of chicken dishes, stews, and sauces, adding a distinct Cuban flavor to dishes made with otherwise basic ingredients.

Ingredients:

- 1 can plain black beans, drained
- 2 tablespoons olive oil
- 3 clovies of garlic, peeled and crushed
- 1 medium onion, peeled and chopped
- 1 mild large green chili, such as anaheim or a green bell pepper
- 3 teaspoons of ground cumin
- 1 bay leaf
- 1/2 teaspoon chili powder

Salt and pepper to taste

Preperation:

1. Heat the oil in a saucepan over medium heat. Gently sauté the garlic, onion, and green chili until they are tender.
2. Add the cumin, bay leaf, and chili powder. Stir until mixed with the vegetables.
3. Add the drained beans to the vegetables and stir. Bring to a boil then reduce the heat to low.
4. Cover and simmer for 5 minutes. Adjust the seasoning with salt and pepper.

Mojo* Sauce

Prep time 5 minutes | Cook time 20 minutes

*Say it like this: "**mo**-ho."
Syllable in bold is the strongest.

Ingredients:

- 1/3 cup olive oil
- 6 cloves of garlic, minced
- 1/3 cup orange juice
- 1/3 cup lime juice
- 1/2 teaspoons of ground cumin

Salt and pepper to taste

ADVENTURE 8

Mojo is best when served within a couple of hours of making, but it will keep for several days, well-capped in a jar or bottle, in the refrigerator. It is often served on boiled yucca, baked plantains, or grilled meats.

Preperation:

1. Heat the olive oil in a deep saucepan over medium heat.
2. Add the garlic and sauté until lightly toasted, which should only be about 20 to 30 seconds!
3. Add the orange and lime juice, cumin, and salt and pepper. Watch out! The sauce may sputter.
4. Bring to a rolling boil. Reduce heat and simmer for 2 to 3 minutes. Taste and correct seasoning, if needed.

    Cool before serving.

Show off your cooking skills!

Have your grown up take a photo, and share on social media using the hashtag:

#KnowYourAdventure

KnowYourselfOAK   KnowYourselfOAK

THE ENDOCRINE SYSTEM

Thoughts for Young Chefs

What did you learn about Cuban food that you didn't know before this adventure?

_____

_____

_____

Know Yourself Adventure Recipes

Review the recipes that you eat everyday.

Record similarities and differences to Cuban food in the Venn Diagram below.

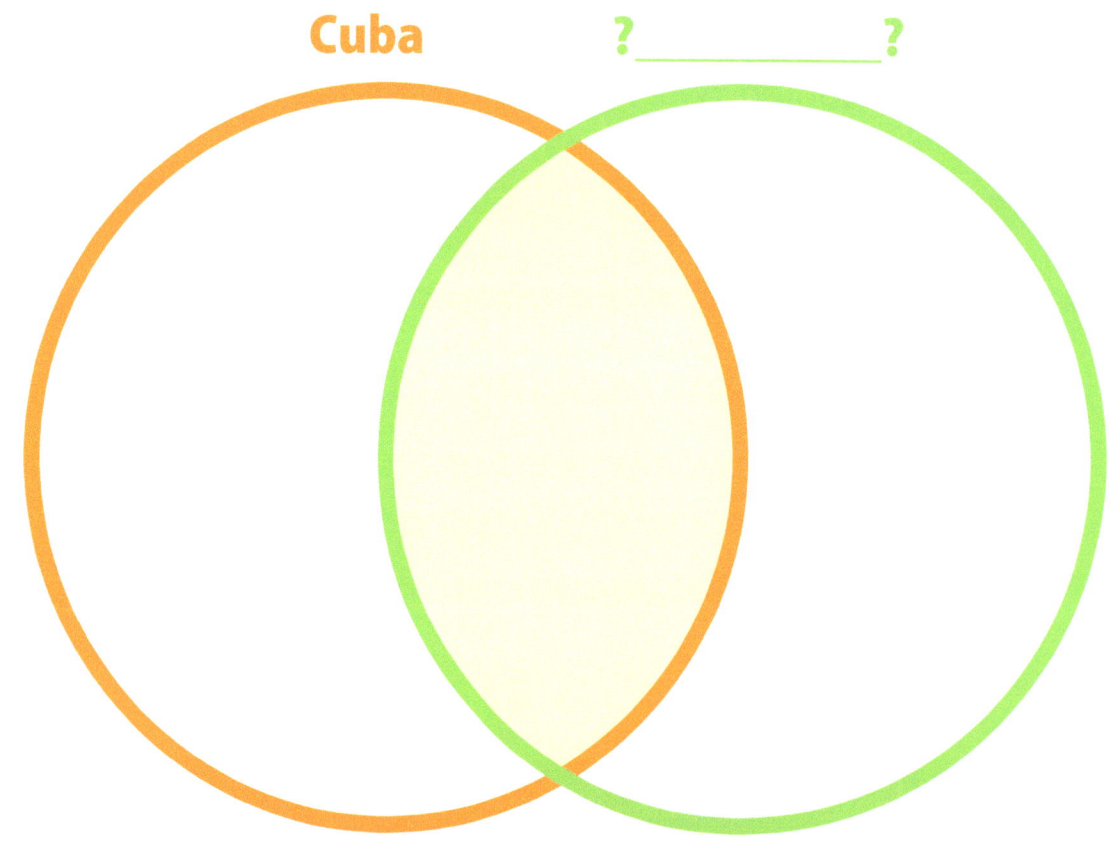

What type of food are you inspired to make next?

_____

_____

_____

_____

_____

_____

_____

Show off your cooking skills!

Have your grown up take a photo, and share on social media using the hashtag:

#KnowYourAdventure

KnowYourselfOAK   KnowYourselfOAK

In It to Win It!

Adventurer, it's been a tough game so far but you've made it to the bottom of the 9th! I think you're ready to win this! To bring it home, answer these final questions.

When Cuba gained independence, many who ran businesses growing sugarcane and producing sugar became very rich while others grew increasingly poor. While many struggled to make things better for all, the government was often corrupt and sided with the wealthy. When Castro took power, many were excited simply for the promise of an honest government that represented the people of Cuba. Most people just wanted to get their fair share, though they disagreed on what that meant.

What is something you think is unfair? How does it make you feel? What do you think could be fixed to make it better, not just for you but for everyone?

_____

_____

_____

_____

_____

_____

_____

_____

_____

In this adventure, we learned about the endocrine system, which involved a lot of teamwork from different gland players. When one gland isn't functioning right, it can cause a host of disorders in the body. We also learned about the communist revolution which took not one, but many freedom fighters supporting the fight for Cuban Independence.

What instances in your life do you see many people creating a successful outcome? Can you imagine what would happen if just one of those people were missing? Do you see how every member has an important role? Describe a scenario in your everyday life and write down your thoughts below.

_____

_____

_____

_____

_____

_____

_____

_____

_____

_____

THE ENDOCRINE SYSTEM

Spot the Difference!

There are 17 differences between these two images. Can you spot them?

THE ENDOCRINE SYSTEM 107

Further Reading

Fiction

Drum Dream Girl: How One Girl's Courage Changed Music.
Margarita Engle

Drum Dream Girl is a fictional story based on the life of Cuban musician Millo Castro Zaldarriaga, a young girl told that women cannot be drummers. Practicing in secret, she perseveres and follow her dreams. The author also includes some information on the real life inspiration.

Engle, Margarita, and Rafael López. 2015. *Drum Dream Girl: How One Girl's Courage Changed Music.* **(Ages 7+)**

90 Miles to Havana.
Enrique Flores- Galbis

A great historical fiction for older readers looking to learn more about Cuba after 1859. The story is set in 1961, during the years after the Cuban Revolution. It follows the main character, Julian and his two brother's early experience in Cuba and new life as immigrants in America.

Flores-Galbis, Enrique. *90 Miles to Havana.* Square Fish, 2012. **(Ages 10+)**

Further Reading

Fiction

Under the Royal Palms: A Childhood in Cuba
Alma Flor Ada

An autobiographical collection of stories about childhood in Cuba, Alma Flor Ada's collection of stories growing up in Cuba provide both a look into life for a child in the city of Camagüey. These stories are excellent for reaching across time and space to help kids connect to Cuban life.

Ada, Alma Flor. 1998. *Under the Royal Palms: A Childhood in Cuba.* New York: Atheneum Books for Young Readers. **(Ages 10+)**

Rady Children's Hospital, San Diego is a great online resource to continue learning about the endocrine system. This site and other children's hospital sites make fun and educational material that helps young learners understand their body's system and illnesses.

https://www.rchsd.org/health-articles/endocrine-system/

THE ENDOCRINE SYSTEM

Loops Crew Part 8: The Right Amount

THE ENDOCRINE SYSTEM

111

THE ENDOCRINE SYSTEM

But it's... ...just the right amount!

Now we can build our skatepark!

Now the real fun begins, Loops Crew!

Because above all things, you trusted in you!

BYE FOR NOW!

THE ENDOCRINE SYSTEM

NEXT
The Respiratory System

Buckle your seat belts because this adventure is on the **USS Wanagari**... your guide is **Lieutenant Muthoni Kiaria**, and the **year is 2216.** YIKES! Rocket to knowledge about the respiratory system in an out-of-this-world adventure in ... **LIVE LUNG AND PROSPER!**

Get to Know...

Stokely

A skateboarding whiz kid who loves all things food, Stokely might be the youngest Time Skater, but don't try and out-trick him on the mini-ramp. Curious about the world and good with tools, he's never too busy to help a friend in need.

Age: 7

Enjoys: Playing basketball, trying new foods, helping others

Favorite Color: Blue

Admires:
Kevin Mazure (actor)
Harry Sheehy (former coach)
Kobe Bryant (basketball player)
Gordon Ramsey (chef)

Favorite Quote:
It is your response to winning and losing that makes you a winner or a loser
— Harry Sheehy

Answer Keys

Evolution of the Revoluion

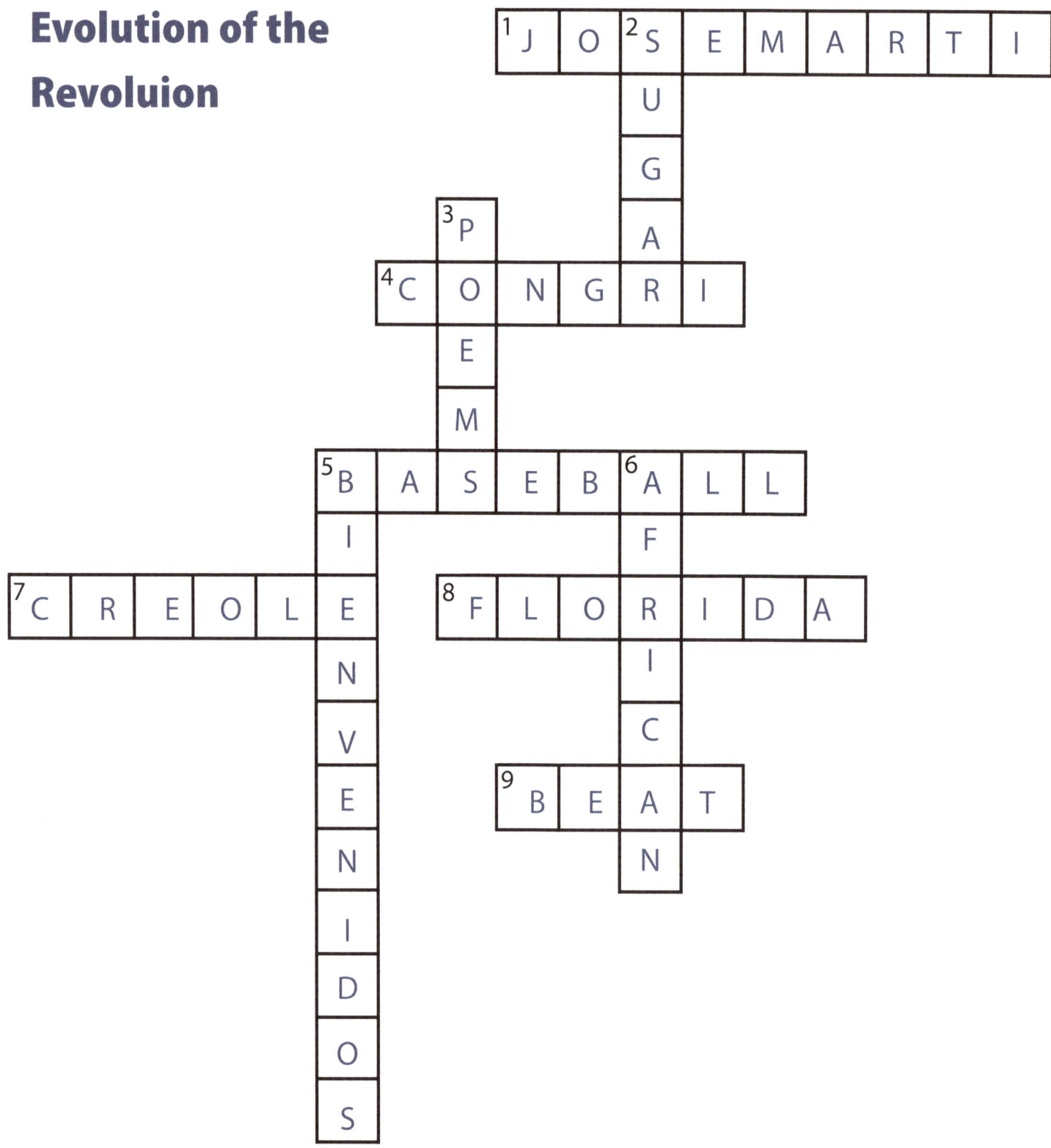

Cuban Comprehension

Havana was established by the SPANISH, though it was briefly captured by the BRITISH. The Spanish brought African slaves to Cuba to harvest SUGARCANE. In 1902 the Republic of Cuba was created, gaining INDEPENDENCE for the Cubans. After becoming independent, the SUGAR industry made many wealthy, but many still lived in POVERTY.

In 1959 the revolutionaries under FIDEL CASTRO took over, trying to create a communist society. The government took control of everything, even banning professional BASEBALL and the board game MONOPOLY. When a million African slaves were brought to Cuba, they brought with them their music and formed the style SON CUBANO. It uses a mix of SPANISH songs with AFRICAN rhythm and percussion. The Cuban instrument of the CLAVE is used in these rhythms. Cuban students brought back the sport of BASEBALL in the 1860's from trips to the U.S. and it became quickly popular. When an attempt to gain independence failed in 1869, the Spanish banned it, trying to replace it with BULLFIGHTING!

This turned it into a symbol of RESISTANCE.

Answer Keys

Meet Your Match Answer Key:

Pancreas - Regulates sugar levels in the blood; Pineal - Secretes the hormone melatonin; Parathyroid - Regulates calcium in your body; Gonads - Makes and secretes hormones that control mood; Thymus - Produces and Secretes the hormone thymosin; Hypothalamus - The "General Manager" of the Endocrine system; Adrenal - Influences your metabolism and stress response; Pituitary - Works with the Hypothalamus to regulate other glands; Thyroid- Regulates your metabolism, growth, and development

Talk About A Team Answer Key:

Your e N D O C R I N E system is responsible for regulating, controling and c O O R D I N A T I N G many bodily functions. It uses h O R M O N E S which act like chemical sparks to communicate between cells, tissues, and o R G A N S. Batting up your endocrine system for the big game (a very important one!) is your h Y P O T H A L A M U S, also known as the "General Manager." The p I T U I T A R Y gland works with the hypothalamus to regulatre other glands. Other gland team players are the t H Y R O I D, a D R E N A L, p I N E A L, p A R A T H Y R O I D, t H Y M U S, p A N C R E A S, and r E P R O D U C T I V E. The reproductive gland can be called g O N A D S, one key difference in genders is that boys have t E S T E S and girls have o V A R I E S.

Answer Keys

Endocrine Expertise Answer Key:

```
W S P R M S H I A F B Z M D G A P C
F T S A E E M Y K E Q P S P C D J G
T P P Q R P D K P R Z B P I X R L C
Y H I N R A R U F O U U H N G E C J
M E Y T H B T O L Z T V J E X N F E
T E B R U W S H D L Z H X A S A M U
J Y Q W O I V U Y U A L A L O L B U
T S J A M I T F G R C M M L G V L T
L E O T E T D A B A O T G J A N J B
T H Y M U S B W R V R I I L J M I Y
C O R T E X W N U Y Y W D V F W U F
W X U P P A N C R E A S Y U E P S S
```

Chemical Composition Answer Key:

1. Hypothalamus
2. Pituitary
3. Thyroid
4. Adrenal
5. Pineal
6. Parathyroid
7. Thymus
8. Pancreas
9. Reproductive

THE ENDOCRINE SYSTEM

CREATED WITH LOVE
BY THE
KNOW YOURSELF TEAM

KnowYourself.com   KnowYourselfOAK   KnowYourselfOAK

www.ingramcontent.com/pod-product-compliance
Lightning Source LLC
LaVergne TN
LVHW070951070426
835507LV00031B/3495

9 780999 881975 4